Addition S[trategies]

by April Barth

Table of Contents

Words to Think About 2
Introduction 4
Chapter 1 Counting On 6
Chapter 2 Making Doubles 8
Chapter 3 Making Ten 10
Conclusion 14
Glossary and Index 16

Words to Think About

count on

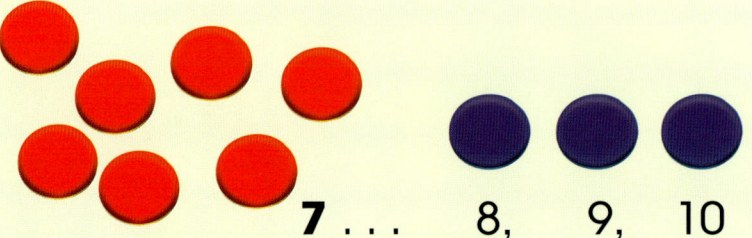

7 ...   8,   9,   10

We count on to find how many in all.

doubles

$$2 + 2 = 4 \quad 3 + 3 = 6$$
$$4 + 4 = 8 \quad 5 + 5 = 10$$

We see four sets of doubles.

part

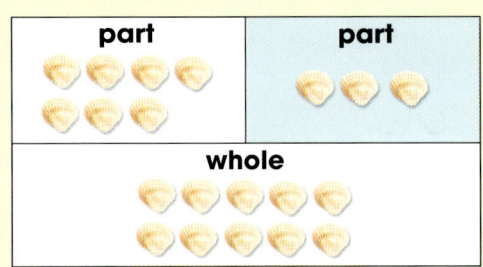

One part has 3 shells.

sum

$$7 + 3 = 10$$

When we add 7 and 3, the sum is 10.

ten

We see ten shells.

whole

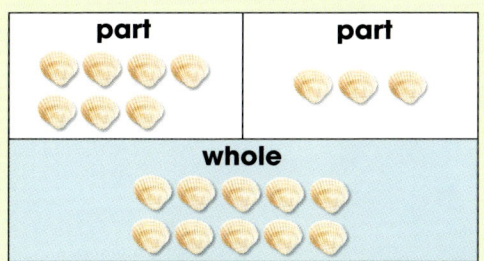

We join two parts to make a whole.

Introduction

Sometimes you need to know how many in all. You can add to find the **sum**.

You can use counters to make models. The models can help you add **parts** to make a **whole**. You can use addition strategies, too.

Chapter 1
Counting On

If you can count, then you can **count on** to find how many in all. Eight turtles hatched from one nest. Three turtles hatched from another nest.

Let's count on to find how many turtles there are in all. Start with 8 and then count 3 more: 8, 9, 10, 11. There are 11 turtles in all.

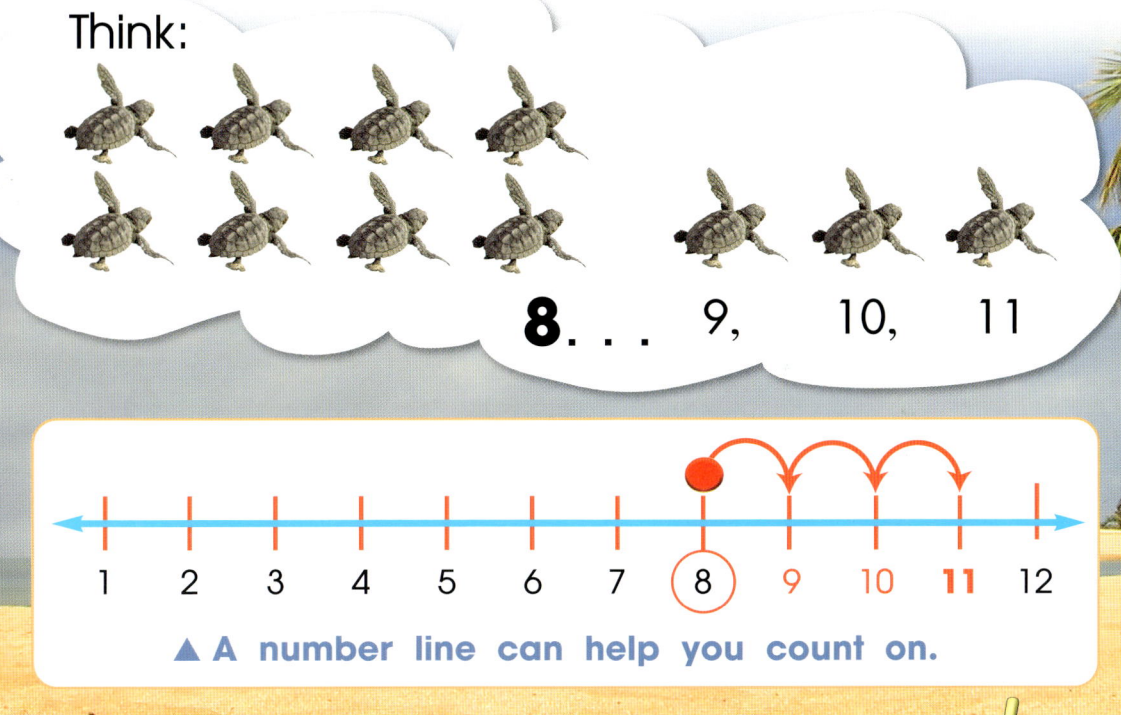

Think:

8. . .   9,   10,   11

▲ A number line can help you count on.

Your Turn

- Choose a number between 5 and 8. Use counters to show that number.
- Now add two more counters.
- How many counters are there in all?
- Tell what strategy you used.

Chapter 2
Making Doubles

If you add a number to itself, then you are making a **double**. Look at the numbers below. Do you see a pattern?

Making doubles is an addition strategy. You can use this strategy to find certain sums. Think about 6 plus 7. What double can you use to help you find the sum?

6 + 7 = ?
6 + 6 + 1 = ?
  └─double─┘

12 + 1 = 13

▲ You can find sums that are doubles plus one.

Your Turn

- First, use cubes to show making doubles. Then add one more cube.
- Use numbers to show what you did.

Chapter 3
Making Ten

Making **ten** is a powerful addition strategy. A ten frame can help you solve problems. How many coconuts are there in all?

$7 + 5 = ?$

▲ A ten frame can help you solve the problem.

5 coconuts

7 coconuts

Imagine filling the ten frame to find the sum. There are 7 brown coconuts. There are 5 green coconuts.

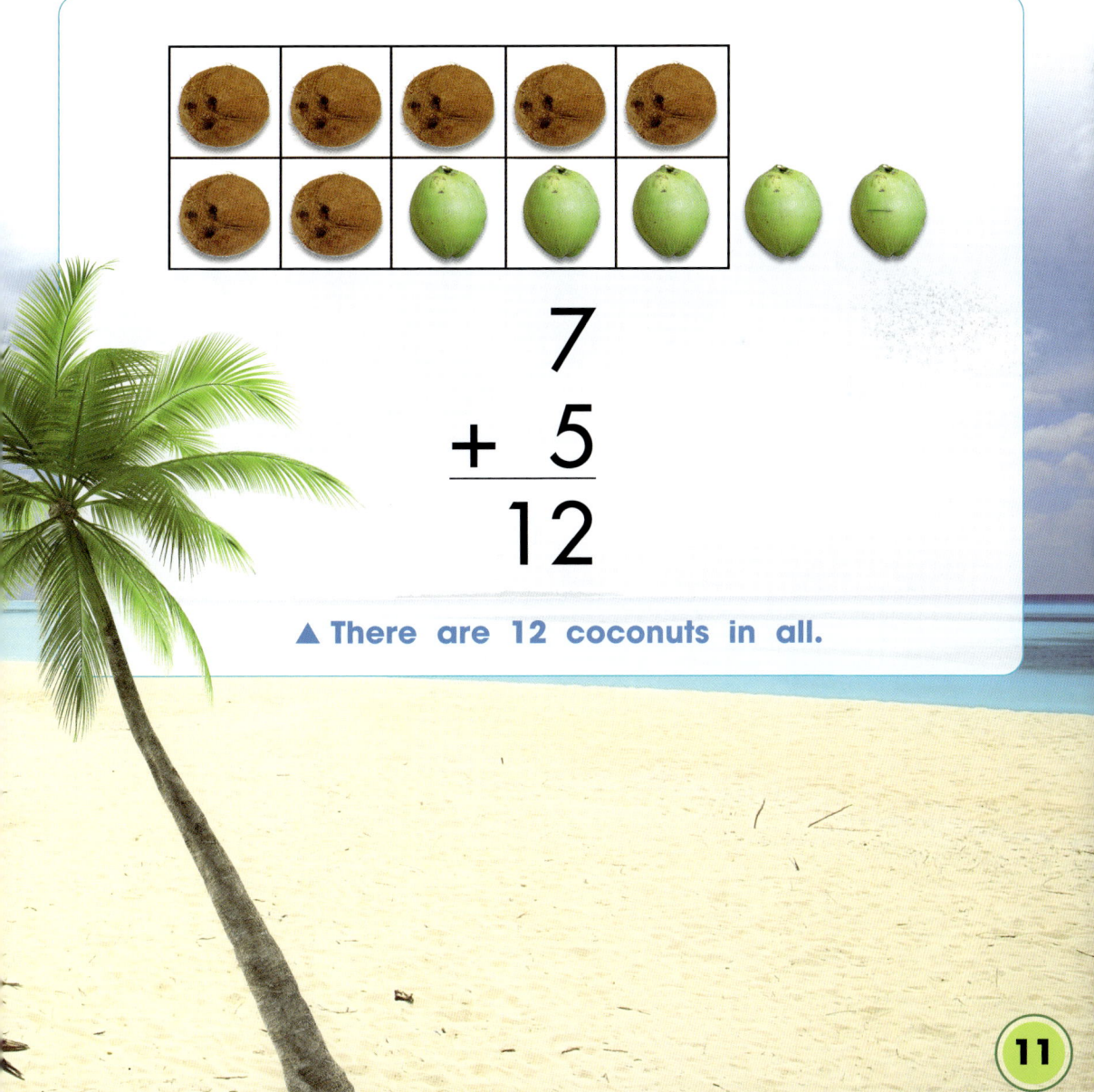

▲ There are 12 coconuts in all.

Chapter 3

Nine yellow fish swim on the reef. Four blue fish are swimming on the reef, too. How many fish are there in all?

9 + 4 = ?

9 + 4 = ?

▲ Make 10. There are 10 and 3 more.
There are 13 fish in all.

Making Ten

Eight sailboats have striped sails. Six sailboats have red sails. How many sailboats are there in all?

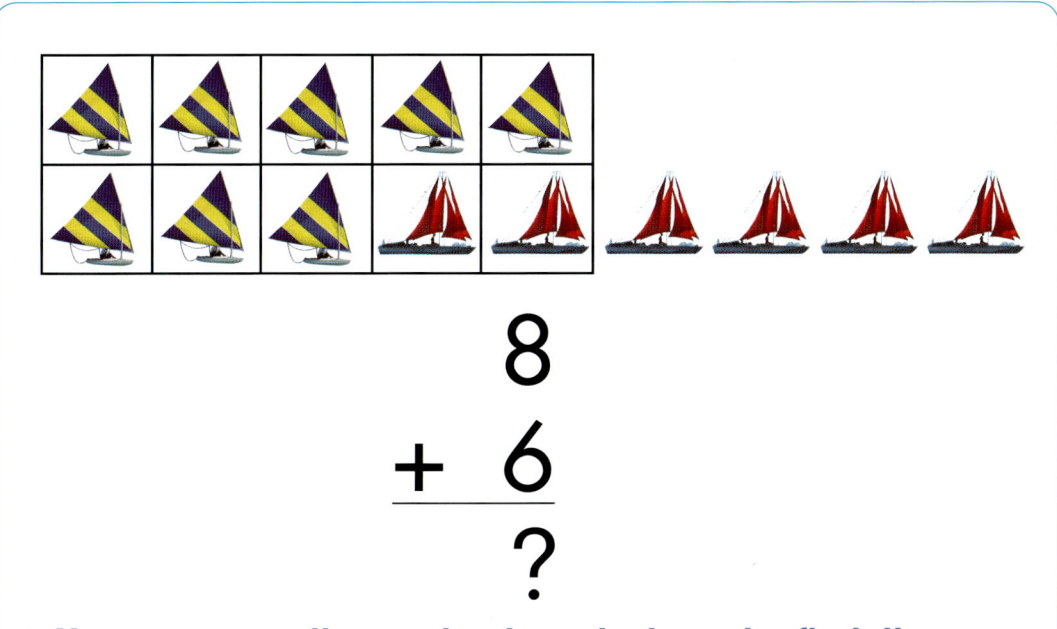

▲ You can use the make ten strategy to find the sum.

Conclusion

You can add! Addition strategies can help you find sums.

Use models.

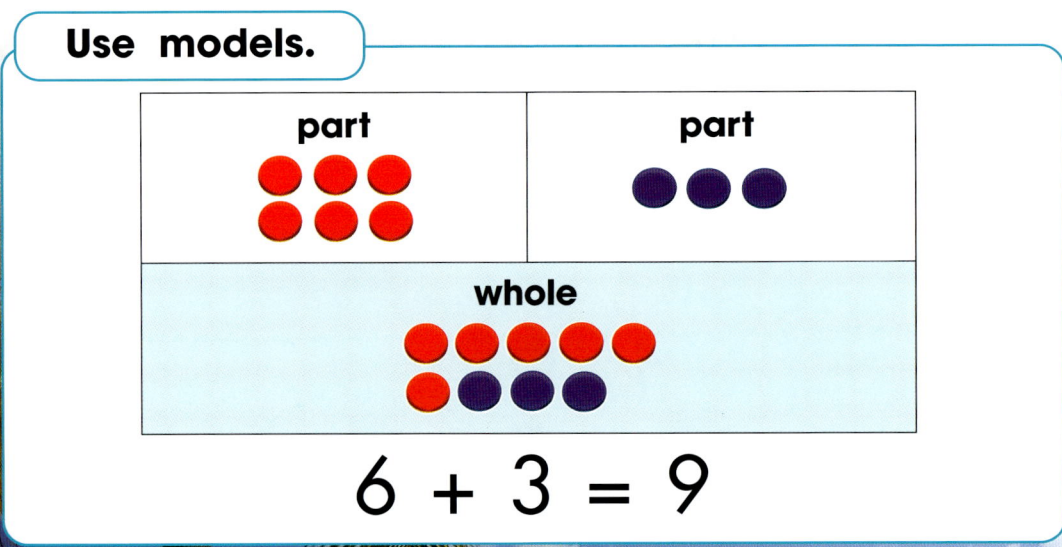

$6 + 3 = 9$

Count on.

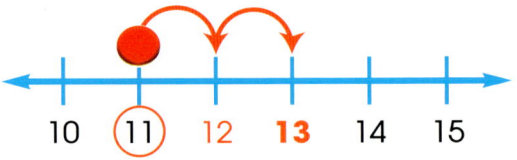

$11 + 2 = 13$

Counting on, making doubles, and making ten are some addition strategies.

Make doubles.

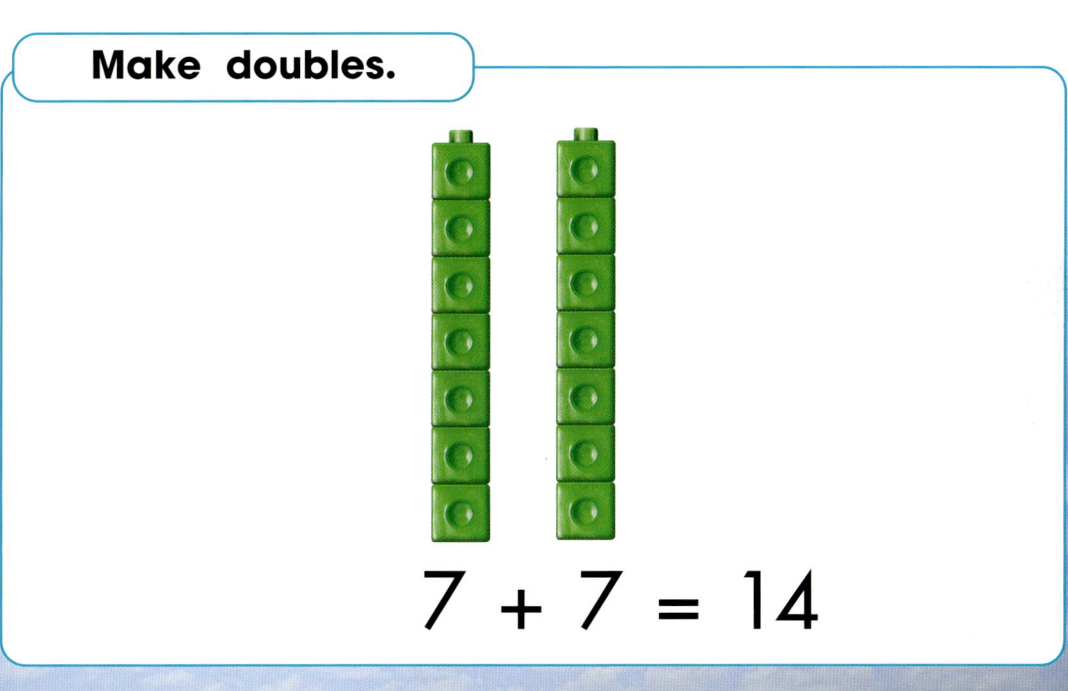

$7 + 7 = 14$

Make ten.

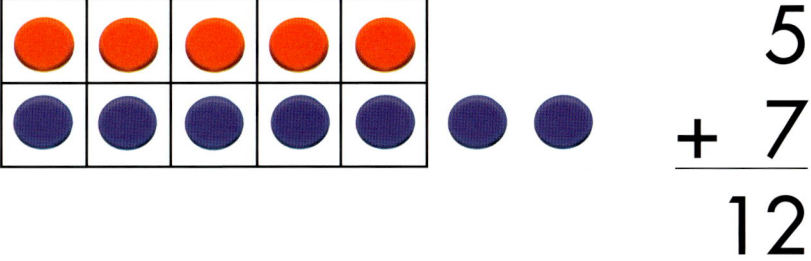

$$\begin{array}{r} 5 \\ + 7 \\ \hline 12 \end{array}$$

Glossary

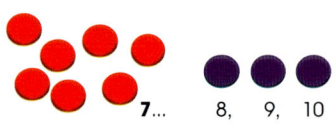

count on a strategy for adding 1, 2, or 3 more

See page 6.

doubles when we add a number to itself

See page 8.

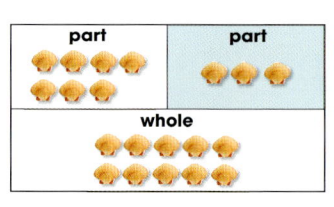

part a piece

See page 5.

7 + 3 = 10

sum total; the result of addition

See page 4.

ten one more than nine and one less than eleven

See page 10.

whole an entire amount

See page 5.

Index

count on, 6–7, 15

doubles, 8–9, 15

part, 5

sum, 4, 9, 11, 14

ten, 10–11, 15

whole, 5

16